PILOT'S NOTES

FOR

MESSERSCHMITT BF109

ISBN 085979119X
780859791199

TECHNICAL DATA AND HANDLING NOTES

Messerschmitt Bf 109 Pilot's Notes

This Pilot's Notes document is based on original archive material in the form of technical data and handling notes for the Bf 109, including a translation from the original German operating instructions.

As far as possible the original wording and syntax of the historical documents has been preserved except in a few cases where wording or punctuation has been changed to aid clarity.

The photographs and diagrams are courtesy of the Public Records Office.

Edited by Paul Sinnott

Messerschmitt Bf 109 Pilot's Notes

Part I – Messerschmitt Bf 109E

Author's Note

The first part of these notes are based on a report produced by the Royal Aircraft Establishment, Farnborough. A captured Messerschmitt Bf 109E was tested during September 1940. The findings also used a report produced by the French who had captured two Bf 109s earlier in the war.

Paul Sinnott, September 2000

Scope of Investigation

The handling tests conducted on the Bf 109 covered the following areas: take off and landing; trim and stability; 'one control' tests, flat turns and sideslips; stalling tests; high speed dive and 'feel' of the aircraft's controls.

Performance of the Aircraft

The weight of the aircraft during these tests was 5,600lb, there is an error factor of 2-3% for the speeds quoted.

Top Level Speed:

At 16,400ft - 355mph, 2,400rpm +2.3lb/square inch boost pressure with radiators closed.

At 14,800ft – 330mph, 2,400rpm +2.1lb/sq.in. boost pressure with radiators open.

Rate of Climb to Height

Height (ft)	R.P.M	Boost lb/sq in	Rate of Climb Ft/min	Measured time to height
3,280	2,430	+3.9	2740	1min – 16sec
6,560	2,430	+3.7	2530	2min – 31sec
9,850	2,430	+3.6	2900	3min – 50sec
13,150	2,430	+3.5	2990	5min – 03sec
16,450	2,430	+2.0	2600	6min – 20sec
19,750	2,400	+0.2	1860	8min – 01sec
23,000	2,400	-1.8	1450	10min – 02 sec
26,300	2,400	-3.6	820	13min – 35sec

Absolute ceiling – 32,000ft.

General Description

The Bf 109 is a low wing monoplane of clean design. The construction is all-metal except for the moveable control surfaces and flaps which are fabric covered. The standard of surface finishing is high, all the metal coverings being flush riveted. The design allows for easy maintenance and all parts can be readily removed and replaced. The all up weight of the aircraft is 5,600lb.

Engine

The engine fitted is a Daimler-Benz DB 601. The engine consists of a 12 cylinder, 60° inverted V, liquid cooled with spar reduction gear. This produces 1,100 b.h.p. There is provision to fit a cannon firing along the axis of the reduction gear through the airscrew hub. Bosch injection pumps meter the fuel.

The engine is glycol cooled with two radiators, one under each wing close to the trailing edge. Each radiator has a flap, operated by a wheel in the cockpit, which can be used to regulate the cooling. An oil cooler is positioned in a duct beneath the fuselage. The air intake for the supercharger is in the form of a scoop protruding from the left side of the front fuselage. The supercharger is driven through a hydraulic coupling.

Part I – Messerschmitt Bf 109E

Airscrew

The airscrew has three metal blades forming 10.2ft diameter and is of variable pitch as fitted on most German aircraft of the time. The pitch is controlled electrically.

The pitch can be set between 22.5° and 90° making the airscrew fully feathering. Alterations are performed by a stationery electric motor fixed to the crankcase immediately behind the airscrew hub, this operates through a flexible drive and differential reduction gear. The cockpit has a pitch indicator which is coupled to the electric motor mechanically. The indicator is in the form of clock face with hour and minute hands – one degree being represented by ten 'minutes' on the clock. The pilot has no provision to govern the r.p.m. and must set the pitch accordingly.

Fuselage

The wing-body junction has a very small fillet, the side of the fuselage is roughly normal to the upper surface of the wing. The fuselage's under surface remains flat, flush with the under surface of the wing for more than 3ft to the rear of the trailing edge. The fuselage cross-section then gradually assumes an ovoid form, finally becoming pear shaped near the tail, the fin is merged gradually into the fuselage.

60lb of permanent ballast is carried at the back of the fuselage – this is to counteract the extra weight of the DB 601 engine compared to the previously fitted Jumo 210 unit.

The undercarriage attaches to the fuselage at the wing root, retracting sideways into the wings. This allows the aircraft to stand on its undercarriage with the wings removed, this easing maintenance.

Wings

The wings have a straight taper in plan, a sweepback of 1° at the quarter chord line. The wing tips are square cut, the ratio of root chord/tip being 2.06:1.

The thickness chord ratio is 0.148 at the root and 0.105 at the tip. All along the span the aerofoil section has a 2% camber with the maximum thickness at about 30% of the chord. There is no wing twist, the dihedral angle at 5.75° is fairly high.

The surface finish is again good featuring flush riveted metal skin, all inspection doors being flush and well-fitting when closed.

Slots

The slots occupy 46.2% of the span and extend well inboard of the ailerons. The slat is articulated to two transmission rods which run straight out of the wing and are linked together by a robust system of rigid rods and bell-crank levers. The slots open and close very freely and when closed fit very well between the wing and slat. There is no damping device in the mechanism.

Flaps

The 25% slotted flaps occupy 51.8% of the wing span. The ratio of flap area to wing area is 0.14, the maximum angle, 42.5°. The portion of slotted flap immediately behind the radiator is thickened in section to prevent the radiator flap from stalling at when fully open at low climbing speeds. The bulge of the flap considerably reduces the very large expansion at the rear of the radiator duct with the radiator flap open.

Control Surfaces

When tested on the ground the control surfaces showed very little friction and no appreciable backlash. Each surface is mass balanced and the control column is statically balanced for fore and aft movement. This is achieved by using a counterweight incorporated into the elevator control circuit. The transmission system between the cockpit controls and control surfaces in by a mixed system of rods and wires.

Ailerons

The slot width is very small, the ailerons have 21.6% balance and the hinge position is well below the control surface. Each aileron has a mass balance weight of streamlined form carried on an arm projecting downwards. The centre of gravity of the aileron coincides with the hinge line.

When the flaps are up the ailerons have a droop of 1.2° and 2:1 differential – maximum angles are 13.5° down and 25° up. As the flaps are lowered both ailerons come down progressively. When the flaps are fully lowered (42.5°) both ailerons come down to 11°, the differential is now slightly less.

A small fixed trimming tab is attached to each aileron. These tabs are of metal sheet and can be bent on the ground to adjust lateral trim.

Tail Unit

The tailplane is high with a large proportion of the rudder coming below it. It thus remains unshielded at high incidences, useful for anti-spinning. No flight operable rudder trimmer is provided. Controls for the fin and rudder are horn balanced; the percentage balance is 10.35% for the elevator and 8.5% for the rudder. The mass balance weights are carried in the horns.

Longitudinal trimming is effected by means of an adjustable tailplane having a 12° incidence range and is operated mechanically by a handwheel to the pilot's left. This wheel is mounted concentrically with the flap-actuating wheel and by winding both wheels together. The pilot automatically compensates for changes of trim due to the flaps.

Flying the Aircraft

Pilot's View of the Cockpit

Size

The cockpit is certainly too cramped for comfort. The width is too narrow to fit in comfortably or to allow movement within the cockpit. The headroom is also insufficient for taller pilots. When wearing a seat fitted parachute a pilot of average height's head will touch the hood roof. The seat position is also tiring when flying for extended periods.

Noise

The cockpit is noisy at full throttle when compared to a Spitfire.

Part I – Messerschmitt Bf 109E

Main Flying Controls

The control column feels good and the slight offset of the grip eases flying. The rudder pedals are positioned such that the pilot leans back contributing to the lack of comfort during long flights. The pedals cannot be adjusted forwards or backwards to accommodate pilots' varying heights.

Trimming and Flap Controls

These are excellently placed to the pilot's left. The flap gear is manual and extremely good being easy to operate and reliable. A noted feature is the juxtaposition of the tailplane adjusting wheel and the flap control wheel. This means the controls can be operated together with one hand, the change of trim flaps are thereby automatically corrected.

Throttle

The throttle arrangement is extremely simple, consisting of one lever with no gate or override. The DB 601 engine's response to throttle opening is very good due to its direct injection system. This makes it impossible to choke the engine and means it does not splutter or cut out when under negative 'g' if the stick is suddenly pushed forward. This is a crucial factor in air-to-air combat.

Airscrew Control

The works will with no difficulty experienced when in use. The pitch control lever is placed on the dashboard.

Undercarriage Control

This works simply and efficiently, it has no warning hooter as fitted on British aircraft.

Brakes

The brakes are foot operated and work well though do not have very sensitive controls.

Instrument Panel

The instruments are well grouped with the flying instruments to the left and the engine instruments to the right. The dials can be clearly read, illumination for night flying is limited. There is no gyro horizon or blind flying panel.

Ancilliary Equipment

The machine gun and cannon firing switches are positioned in the grip of the stick and are well placed. The electrical panel is to the lower right of the dashboard. The buttons on it are difficult to distinguish for a pilot not experienced on the aircraft. There is a jettison arrangement for the Verey cartridges to allow the pilot to quickly dispose of them in a forced landing. This would prevent the signal of the day being revealed to the enemy.

View

The flat front panel is inclined at 55° to the horizontal when at flying attitude. The larger corner panels are also flat. The left corner panel is divided into two vertical sections. The forward position hinges forward about its leading edge.

For access to the cockpit the hood is hinged on the starboard side. Sliding windows are fitted, one in each of the hood's side panels. For emergency ejection the hood is spring loaded. When the jettison leaver is pushed the whole hood and radio mast are flung clear backwards.

The view when taxiing is poor. This is due to high ground attitude of the aircraft and because the hood cannot be slid backwards to allow the pilot to look out to one side.

When in flight the view is reasonably good. The struts between the glass panel are thin and do not impair vision, catch the pilot's eye or create blind spots.

Lateral and rear vision is fairly good but due to the cramped nature of the cockpit it is difficult to look downward or upwards and behind.

The direct vision opening gives a large field of vision and is completely draft free at all speeds. This allows for high flying speeds to be maintained during heavy rain. The opening also assists landing when the high nose position obstructs forward vision.

The windscreen panels are clear and free from distortion. Oiling up in flight is not a problem. The hood sliding panels are difficult to open particularly at high speeds.

Flight Test

Take-Off

The slotted flaps are set at the recommended position of 20° for take-off. The throttle opens very quickly, responding almost instantaneously to throttle movement. The direct injection system means there is no risk of choking. The initial acceleration is very good and the aircraft will not bucket or swing. When running along the ground the aircraft will rock slightly from side to side but the movement is minimal and should not concern the pilot.

On opening the throttle the stick should be held hard forward. The tail will come up quickly and the stick can then be eased back. The pilot should hold the aircraft on the ground for a short time after flying speed has been reached. If the aeroplane pulls off too soon the left wing will not lift. On applying opposite aileron the wing will come up, fall again with the ailerons snatching somewhat. As long as the aircraft is not pulled off too quickly take-off is fairly easy and straightforward.

The take-off run required is very short and initial rate of climb is fast.

Approach

The stalling speed when gliding, with flaps and undercarriage up, is 75mph. With flaps and undercarriage down it is 61mph. Lowering the flaps makes the ailerons heavy and less effective, giving rise to a pronounced nose-down pitching movement. Using the juxtaposing flap and tailplane adjustment operating wheels can easily rectify this. The attitude of the aircraft at constant airspeed changes by roughly 10° when the flaps are put down. Lowering the undercarriage causes slight nose-heaviness.

If the pilot needs to go round again following a failed landing attempt and opens the engine with the flaps down the aircraft will become slightly tail heavy. The aircraft can be held with one hand as the trim is adjusted.

The following table shows the effect of varying the speed of approach as measured when gliding with flaps and undercarriage down. The attitude at cruising level was 5°.

A.S.I. mph	Vi mph	Rate of Descent ft/sec	Gliding Angle	Gliding Attitude
60	80	26.5	13.0°	+3°
70	86	22.5	10.3°	0°
80	93	23.3	9.8°	-5°
90	100	24.5	9.6°	-12°
100	108	29.2	10.5°	-18°

The normal approach speed is 90mph. At speeds below 80mph the pilot will gain the impression of sinking, at above 100mph the impression of diving.

The glide path down is fairly steep at 90mph; the view is good due to the nose-down attitude. Longitudinal stability is extremely good when flown stick free. The elevator is responsive and heavy at this speed. These features make the ease of approach very good. Ailerons can be lowered to 11° with the flaps without reducing their effectiveness, they will feel much heavier. The rudder is sluggish for small movements.

Normal gliding turns can be performed at 90mph flaps down without any stalling or undue loss of height.

Landing

Landing the Bf 109 is difficult. This is due to two main factors. Firstly the high ground attitude of the aircraft and secondly due to the narrowness of the main undercarriage.

The aircraft must be rotated through a large angle before touchdown requiring a fair amount of skill from the pilot, tempting him to do a wheel landing. If a wheel landing is attempted there is a strong tendency for the left wing to drop just before touchdown. When the ailerons are used quickly to bring the wing up they tend to snatch a little resulting in over-correction. To achieve a three wheel landing the pilot should hold off a little high allowing the aircraft to slowly sink on to the wheels. This eliminates the tendency for the wing to drop. After a few practice landings the pilot should become easily accustomed to the landing technique and have no further problems.

The centre of gravity is unusually far behind the main wheels and the brakes can be fully applied immediately after touchdown, without lifting the tail. The ground run is very short with no tendency to swing or bucket. As the large ground attitude causes the nose of the aircraft to be very high visibility is poor for taxiing. Landing at night is very difficult.

Ground Handling

The Bf 109 can be taxied very fast as the large tail weight prevents it from bouncing or bucketing. Turning rapidly is difficult and a large amount of throttle is required combined with firm use of the differential brakes if attempting to manoeuvre in a tight space. Apart from this turning the ground handling qualities are good. The brakes are powerful and can be used without fear of lifting the tail. The brakes are foot operated.

Trim

Lateral Trim

Each aileron has a small fitted tab that can be easily adjusted when on the ground to correct any imbalance between the wings. The ailerons cannot be trimmed during flight. Speed and throttle settings have no major effect on lateral trim provided the plane is flown without sideslip. The rudder has no trimmer so sideslip quite probable, especially at high speeds when the rudder becomes heavy. Because of the large wing dihedral any inadvertent sideslip produces a pronounced rolling movement. This needs to be corrected by the ailerons.

Directional Trim

There is a large variance in directional trim at different speeds. The absence of a rudder trimmer compounds this problem. When flying at full throttle there is very rapid variation in directional trim with speed. The rudder is light at lower speeds and can be easily held at 5° when climbing at 150mph. However at high speeds the lack of a rudder trimmer becomes a significant inconvenience to the pilot. At 215mph the aeroplane is trimmed directionally with no rudder applied. At higher speeds left rudder needs to be applied, at 2° when at 300mph which requires a large amount of force from the pilot. This can become tiring and reduce the pilot's ability to perform high speed left turns.

Longitudinal Trim

An 11.7" diameter wheel on the pilot's left controls the adjustable tailplane. To move the tailplane through the full angular range, +3.4° to –8.4°, five and three quarter turns are required. Winding the wheel forward pushes the nose down.

The nose down change of trim due to lowering the flaps and undercarriage is large, but readily corrected. The Bf 109 is very stable when gliding at low speeds with flaps down. This reduces manoeuvrability in the looping plane and contributes to the heavy feel of the elevator at high speeds.

'One Control' Tests, Flat Turns and Sideslips

The following tables show the relative degrees of static directional stability and lateral stability. When testing to produce these results the aircraft was trimmed longitudinally to fly straight and level at 230mph at 10,000ft at 2,200rpm. In these conditions slight pressure on the left rudder is required to maintain keep the plane flying straight with no sideslip.

	Directional (fixed ailerons)	Lateral (fixed rudder)
Sudden application and release of one control with the other fixed	When the rudder is suddenly displaced through half its maximum travel the plane swings through about 8° in yaw and banks 5° in the direction applied to the rudder and the nose pitches downwards slightly On releasing the rudder the nose swings back quickly and the plane does an oscillation in yaw and roll, which fades out. During this oscillation the aileron control moves from side to side. With left rudder the left wing slowly comes up and the aircraft enters a right hand spiral. With right rudder the wing stays down and the plane turns to the right with gradually increasing bank.	When the ailerons are suddenly displaced the plane banks with no opposite yaw. When the stick is released it immediately returns to the central position. When moved left the left wing rises slowly and the aeroplane banks and turns to the right. When right aileron is applied the right wing stays down and the bank gradually increases. When the ailerons are released the plane is very sensitive to rudder position. Using the rudder on its own raises the wing.
Steady banked turn with one control, the other being held fixed.	Rudder alone can produce good banked turns when the ailerons are fixed centrally. Quarter rudder is required to initiate the turn and very little sideslip results. To recover from the turn half rudder should be applied. If the rudder is applied harshly more sideslip is produced on entry and recovery. This also produces a pronounced nose-down pitching moment which has to be corrected by pulling the stick back. If the rudder is released during a 30° banked turn to left the plane rolls slowly to the right. When rudder is released in a 30° banked right turn the result is an increase in bank and tightening of the turn.	Ailerons alone can produce excellent banked turns. Very little sideslip is produced when entering or leaving the turn, even if the ailerons are used harshly.

Flat Turns and Sideslips

Undercarriage Up:

Flown at 230mph at 10,000ft

Only half rudder was used for the test. Full rudder can be applied using considerable force. This produces considerable nose-down pitching movement so that excessive pull has to be applied to the stick to hold the nose up.

Steady Sideslip when Gliding, Flaps & Undercarriage up

Gliding at 100mph:

The maximum bank angle in a straight sideslip is 5°. Rudder is the limiting control, quarter aileron should be used to hold the wing down against full rudder. The aeroplane is nose heavy during side-slip. It also vibrates and will feel a little unsteady, requiring occasional reaction from the controls to keep speed constant.

If all controls are released during side-slip the nose falls and the plane will swing smartly into the sideslip path and bank will decrease very quickly. The plane will then glide straight with level wings and the nose will slowly rise to the trimmed position and oscillate fractionally in pitch.

Steady Sideslip when Gliding, Flaps & Undercarriage Down

Gliding at 90mph:

The maximum bank angle is again 5°, with rudder the limiting control. One fifth aileron is required in conjunction with full opposite rudder. The aircraft is less nose-heavy with flaps up. Vibration again occurs and the plane will feel unsteady.

When all controls are released the same effect occurs as above but with increased pitching oscillation which also lasts longer.

Stalling Tests

Stalling Speeds and CL Max

Condition		Pilot's A.S.I. mph		Indicated Airspeed mph		CL	
Flaps & Ailerons	Under-carriage	Speed at which slots open	Stalling Speed	Speed at which slots open	Stalling Speed	CL at which slots open	CL Max.
Up	Up	111	754	120.5	95.5	0.865	1.4
Down: Flaps 42.5° Ailerons 10°	Up	90	61	100.5	81	1.2	1.9
Down: Flaps 42.5° Ailerons 10°	Down	90	61	100.5	81	1.2	1.9

High Speed Dives

The plane was put into a 370mph dive, then all three controls were in turn displaced slightly and then released. No vibration, flutter of snaking develops.

With elevator trimmed for normal level flight at full throttle a fairly large push is needed to hold the plane in a dive. If the trimmer is wound forward recovery from the dive is very difficult due to the excessive heaviness of the elevator at high speeds.

When 370mph is reached considerable pressure is needed on the left rudder bar to keep the plane straight. If the rudder is released the plane will bank and turn to the right. Small rudder movements, sufficient for 10° yaw will not produce a pitching movement. Large rudder displacements will cause the nose to pitch down.

Flying Controls

Ailerons

Aileron control is very good at low speeds, there is positive feel and a definite resistance to stick movement.

As speed rises the ailerons gradually become heavier but retain excellent response, being at their best between 150 and 200mph. Between 200 and 300mph they become noticeably heavier and over 300mph become too heavy for comfortable manoeuvring. Over 400mph the pilot can exert up to one fifth aileron using all his strength.

Throttling back will not alter the effectiveness of ailerons at any speed. At low speeds lowering the flaps (ailerons come down 11° with the flaps) results in the ailerons becoming much heavier and slightly reduces their effectiveness.

The ailerons can snatch as the wing tip slots open. This is pronounced during manoeuvring. For instance is the stick is pulled back during a tight turn, producing additional 'g', the slots will open at high airspeeds. As they open the stick will snatch laterally through several inches in either direction. This snatch is sufficient to spoil a pilot's aim during a dogfight.

Some snatching will also occur when gliding close to stall speeds with flaps up and slots open. This will disappear when the flaps are fully lowered, so shouldn't concern the pilot during the approach glide.

Elevator

The elevator gives exceptionally good control as low speeds. It is fairly heavy and not too sensitive during the approach glide with excellent response. Over 250mph the aileron becomes much heavier and between 300 and 400mph restricts manoeuvrability during looping.

At low speeds the elevator is slightly lighter if the engine is throttled back and slightly less responsive. Lowering the flaps will not effect the elevator control.

Rudder

The rudder is light but sluggish at slow speeds, large displacements will be needed to produce quick response. As speed increases so will rudder response. Between 200 and 300mph the rudder is the lightest of all the controls for small movements so should be used for directional aim during dog-fights. Over 300mph the lack of a directional trimmer will be felt as a small amount of left rudder needs to be applied to prevent side-slip. Constantly applying this rudder can cause the pilot to become tired.

Throttling back at slow speeds will make the rudder a little more sluggish. Lowering the flaps further reduces rudder effectiveness though response is still adequate with larger displacements.

If speed is increased while gliding with flaps down the rudder will start juddering slightly at 100mph. The juddering increases rapidly as speed rises and makes increasing speed to over 120mph inadvisable. The normal approach speed with flaps down should be 90mph, rudder vibration is not noticed at this speed.

Harmony

Between 150 and 250mph the controls are well harmonised, the elevator being the heaviest. At low speeds the sluggishness of the rudder adversely affects harmony. At very high speeds the heaviness of the ailerons mars harmony.

General Flying Impression

Test pilots particularly liked the positive feel of the Bf 109's controls at low and medium speeds. Features they disliked were the undue stiffening of the controls at high speeds, particularly the ailerons. Also the aileron snatching caused by slots opening while diving. The lack of a rudder trimmer was also considered a major shortcoming.

Aerobatics

Loops must be started from around 280mph. Slots can open near the top of the loop resulting in aileron snatching and loss of direction. This impedes accurate looping.

Rolls can be performed very easily at speeds below 250mph when the ailerons are light and very effective. There is a tendency for the nose to fall at the final stages of a roll, meaning the stick must be pulled well back in order to keep the nose up. Upward rolls are more difficult. At high speeds the heaviness of the elevator means only a gentle pull-out from a dive is possible.

Messershcmmitt Bf 109 vs Spitfire I

Comparative Aileron Characteristics

	Bf 109E	**Spitfire Mk.I**
Maximum Sideways Force average pilot can apply to stick	40	60
Time to 45° bank when applied at 400mph	4	4
Corresponding Aileron Displacement	1/5	1/5
Kb2 at 400mph	-0.145	-0.14
Wing Span ft.	32.4	37.0
Aileron Type	Slotted	Frise
% Balance (area ahead hinge/ total area)	21.6	27.5
Total aileron area/gross wing area	0.0655	0.078
Max. stick travel in.	+/- 4	+/- 8
Max. aileron angles Up	25°	25°
Down	13.5°	19°

Turns at Minimum Radius without Height Loss

Both aeroplanes were tested at full throttle at 12,000ft altitude.

	Bf 109E	**Spitfire Mk I**
Minimum Radius of turn without loss of height, feet	885	696
Corresponding time to turn through 360°, seconds	25	19
Indicated Airspeed Vi, mph	129	133
Pilot's A.S.I., mph	118	126
'G'	2.10	2.65
Angle of Bank	62°	68°

Part I – Messerschmitt Bf 109E

Tester's Comments

The strengths and weaknesses of the Bf 109E were summarised as follows:

Strengths:

i) High top speed and excellent rate of climb

ii) Good control at low speeds

iii) Gentle stall, even under 'g'

iv) Engine does not cut immediately under negative 'g'

Weaknesses:

i) Controls, particularly the ailerons, are heavy at high speeds

ii) High wing loading causes the aeroplane to stall under 'g' and gives it relatively wide turning circle

iii) Aileron snatching occurs as the slots open

iv) Fast manoeuvres at high speed are difficult

v) Absence of a rudder trimmer curtails stability to bank left at high speeds

vi) Cockpit too cramped for comfort during combat

The gentle stall and good control under 'g' enable the pilot to get the most from the aircraft during dog-fighting by flying very near to the stall. This allows the Bf 109 to remain on the tail of a Spitfire. Both the Bf 109 and Spitfire suffer from the inability to roll fast at high speeds because of the heavy ailerons.

Handling Summary

i) Take-off is fairly easy and straightforward. The controls have excellent feel and the plane is very stable in a glide. Landing can be difficult for pilots unused to the aircraft. Taxiing characteristics are good.

ii) Absence of a rudder trimmer is a handicap at high speeds as there is a large variance in directional trim as speed changes. Longitudinally the aircraft is very stable.

iii) Good banked turns can be achieved using ailerons or rudder alone. Sideslip produces a large nose-down pitching moment.

iv) The stall is not violent and there is no tendency to spin. CL maximum when gliding is 1.4 with flaps up. 1.9 with flaps down.

v) There is no vibrating or snaking when controls are displaced slightly during a high-speed dive.

vi) Aileron snatching occurs when the slots are open. All three controls are heavy at high speeds.

Messerschmitt Bf 109 Pilot's Notes

Part II

Operating Instructions for the Messerschmitt Bf 109T

Author's Note

The following operational instructions are based on a translation of a German document captured by British Air Intelligence in July 1941. The instructions bore the designation 'L.Dv.T.2109 T-2 Fl "Bf.109T-2 Operational Instructions F1, Operation and Maintenance of the Aircraft, July 1941"', Air Minister for the Reich and Chief of Air Force Department of Maintenance and Supply, JA Udet.

The Bf 109T was designed as a sea version of the Bf 109E-3 by Fiesler for the never completed German Navy aircraft carrier *Graf Zeppelin*. This variant was later further developed by Blohm und Voss as the Bv 155 high altitude fighter. The Bf 109 T-2s completed served with 11/JG11 and 1/JG77 and the Heligoland fighter squadron.

1. Getting Ready for Take-Off

A. Towing

The eyelets are on the axle bars. Length of each towing cable should be at least 6m. Aircraft may also be towed by means of a shaft attached to the tail-skid.

N.B.

Do not pull at one eyelet only

Do not pull at the slots

Do not push at the ailerons, landing flaps and horizontal and vertical tail surfaces

B. Fuelling

1. Fuel

Grade of Fuel: Leaded fuel C3 (octane number 100)

Tank Capacity: 400 litres

Filling hole under inspection hole cover on left side of fuselage behind cabin. On putting on the outboard attachment cover see that the pin of the ventilation valve does not remain hanging in pressed down position.

Measurement of Supply:

By means of measuring (immersion) rod (aircraft in tail-down position). Measuring rod is stored beside main tank head under hinged cover plate behind pilot's seat.

By means of electric gauge (aircraft in horizontal position). Indicator on instrument board. Switch for switching on measuring installation on main switch panel to the right.

N.B. Whilst fuelling up, do not connect up electric wiring system.

2. Injection Fuel

Grade of Fuel: 95% petrol and 5% oil

In cold weather 50% petrol, 45% ether and 5% oil

Tank Capacity: 200.5 litres, Filling hole (on tank) in pilot's cabin to the right.

3. Lubricant

Grade of Lubricant:

Stavano 120 (for summer and winter). Tank capacity: 29.5 litres available for lubricant, leaving 6.5 litres as air space.

Filling Hole: Under inspection hole cover in left fuselage wall in upper forward position. Fill up to maximum position in accordance with measuring rod, using funnel with hair sieve.

Measurement of Supply

By means of measuring (immersion) rod. Measuring rod is stored under inspection hole in left upper rear engine cowling.

4. Coolant

Grade of Coolant

Pure soft water. Add anti-corrosion oil 39. Observe strictly instructions on Notice Sheets No.I/125 and I/126. Total capacity of coolant system approximately 60 litres. Filling hole under inspection hole cover in left side of the forward upper portion of the engine cowling.

N.B. When engine is hot, open filling screw cap carefully. (Danger of violent ejection of hot water).>

When filling, aircraft must be in tail-down position. Fill with watering –can to overflowing. Fill up with cold water only when engine has cooled down.

In winter and summer add 50 parts by volume of glycol (F1 No.44 440) to the water. Replenish with pure water when losses occur in summer.

5. Oxygen for Breathing Apparatus

Filling hole: Under inspection hole cover on right fuselage wall.

To fill:

a) Open remote control valve (on apparatus box on right side wall of pilot's cabin)

b) Fill up with oxygen until pressure difference is equalised, then with filling pump increase pressure to 150 at excess (pressure gauge on apparatus box).

c) Close remote control valve

N.B. Do not mix up oxygen with pressure air. See that attachment is absolutely free of oil. Danger of explosion, see L.Dv.291.

C. Checking Up before Starting

See –

1) that instructions of loading plan have been duly observed

2) that the surface of radiator and oil cooler fins are clean; if necessary apply strong water jet in direction from rear to front onto radiator (situated under wing surfaces)

3) that metal and fabric covering are in proper condition

4) that all inspection hole covers, flaps and snap locking devices fit tightly

5) that no loose parts lie about the aircraft

6) that control rods respond on moving the hand lever

7) that controls move easily without excessive play in the bearings

8) that there are no pipe leakages

9) that lubricant filter is clean by repeated pull on the handle situated on the lower engine cowling behind the lubricant radiator

10) that the aircraft is free of ice, grime or snow

D. Starting up of Engine

N.B. For cold-start, observe H. 'Cold Start Preparation'.

1) Close cabin (Cabin handle in position 'closed')

2) Undercarriage switch should be in position 'Rest'.

3) Connect up electric wire system (pull knob on right behind pilot's head)

4) Main fuel cock lever should be in position 'open'.

5) Open throttle to about one third.

6) By means of hand fuel pump fill up fuel pipes.

7) Give instructions for winding up inertia starter (cranking shaft on engine to the right above, starting handle in luggage space).

8) After receiving the clear signal insert key in ignition lock and switch over to M1 and M2.

9) Pull starter knob and when the engine is cold inject at the same time with starting injection pump two to three strokes of injection fuel into the induction air pipe. After working the pump the operating knob must be in contact with the guide nut. On the firing of the engine, let go of the starter immediately.

10) Pull ignition timing handle and hold until engine runs smoothly (this ensures plugs burning clean).

11) After firing, pull throttle lever back to "Idling" until engine runs at 400-500rpm.

12) Watch lubricant pressure: When the engine is cold a pressure of about 6-8 kg/cm2 must be built up within the first 10-15 seconds after firing otherwise stop engine immediately. Test lubricant system and attachments for tightness.

13) Fuel pressure gauge must register 1.3 – 1.5 kg/cm2 within a few seconds after firing, otherwise stop engine. Test fuel system for tightness.

14) After firing of engine strap up starting handle in luggage space and close flap according to instructions.

When the engine does not fire: Wait a few minutes before making the next starting attempt (until inertia starter is completely stopped), then repeat starting procedure. For correcting engine faults see "Operation and Maintenance Handbook for Mercedes-Benz Aero Engine DB 601".

E. Warming Up

N.B. "Cold Start Preparation" H. must be observed when cold starting takes place.

1) Open radiator flaps as required (crank on right beside seat)

2) In cold weather close oil radiator flap (lever on left beside seat)

3) Keep at idling until lubricant pressure has fallen to 4.6 kg/cm2, then increase rpm slowly: the lubricant pressure should never go over 6 kg/cm2.

4) Increase rpm slowly up to a lubricant pressure intake temperature of about 30°C so that lubricant pressure does not exceed 6 kg/m2. During the warming up there should be no sudden rise in lubricant temperature or sudden fall in lubricant pressure, otherwise stop engine and check lubricant circulation.

5) After obtaining the working temperatures, open radiator flaps.

6) Keep under continuous observation power plant indicating instruments.

Test during Warming Up

a) Test ignition system. Switch on ignition switch to "M2" and alternately. If on switching over, the engine speed falls by more than 35rpm, burn the sparking plugs clean at 1,300 – 1,900 by operating the special ignition retarding device.

b) Test fuel system: set main fuel cock lever at about 2,000rpm alternately to 'P2' and 'P1'. Keep in each position for at least _ minute. There must be no drop in rpm, no unsteady running of engine or violent fluctuations in fuel pressure. The fuel pressure gauge should register 1.3 – 1.5 kg/cm2.

c) Test electric installation: In order to ascertain whether generator is supplying current, disconnect electric wire system at approx. 1,900rpm (disconnecting switch on instrument board), and switch on electric instruments – e.g. lamps. The lamps must light up. After the test, switch off electric instruments and connect up again electric wire system (pull knob on right behind pilot's seat).

F. Running up to Power

N.B. For cold start, observe H. "Cold Start Preparation".

To safeguard engine avoid unnecessary running up of engine. Trim tail heavy. Set aircraft exactly nose to wind; if possible load tail.

1) See that cabin is firmly closed.

2) See that blade setting automatic system of airscrew is switched on (tumbler switch on left of fuselage wall: automatic switch on switch panel).

3) Open radiator flaps. If oil temperature is low, keep oil cooler flap closed for the present.

4) Only run up engine when the lubricant intake temperature is at least 30°C (press knob on indicator) and when the coolant outlet temperature is not more than 70-80°C at the beginning of the run-up.

5) Before increasing rpm pull control column towards you and hold in position.

6) Open throttle uniformly to full. The lubricant pressure should not, however, exceed 6 kg/cm2.

7) Thus the following values must be obtained:

Engine speed (at 12 o'clock airscrew setting)	2,300rpm
Induction pressure	1.35 atm.abs
Lubricant pressure	approx. 3.5 kg.cm2
Coolant Temperature	max. 94°C
Lubricant intake temperature	min. 30°C
Fuel pressure	1.3-1.5 kg/cm2

8) Test again briefly ignition and fuel system.

9) Note whether engine runs smoothly and free of vibration.

10) Throttle down slowly.

N.B. When temperature of coolant exceeds 94°C, throttle down immediately to about 800rpm until the temperature falls.

G. Stopping

1) Open radiator flaps.

2) Let engine cool down slowly (throttle down to 800rpm)

3) Keep engine idling for about two minutes, then switch off ignition; withdraw key.

4) Close main fuel cock and whilst the engine is running down pull handle of rapid stop device until engine starts.

5) Press switch disconnecting electric wiring system.

H. Cold Start Preparation

Mixture of fuel and lubricant is effected by means of the mixture valve. Mix only when engine and lubricant will have reached at the next take-off a temperature below approximately +10°C.

Winter operation: below about +5°C air temperature

Summer operation: over about +5°C air temperature

Mixture Table

	Winter			Summer		
Flying time	**Lubricant litres**	**Opening Valve Min.**	**Time of Sec.**	**Lubricant litres**	**Opening Valve Min**	**Time of Sec.**
Over 1 hour	22	3	20	26	1	45
3/4 – 1 hour	23	3	-	26	1	30
1/2 – 3/4 hour	24	2	30	27	1	15
1/4 – 1/2 hour	25	2	-	27	1	-
Under 1/4 hour	26	1	15	28	0	30

Instructions for the Starting Up and Operation of Engine prepared for Cold Starting

1) In very cold weather turn engine by hand at the airscrew with the ignition switched off. Guard against unpremeditated start. In this connection inject by means of the atomiser pump several strokes of starting fuel.
2) Start up normal. At the temperature below –10°C, wind up starter once or rather 2-3 times at only approximately half engine speed and couple in with ignition switched off. Only subsequently wind up inertia starter fully and then start up normal.
3) After firing, watch oil pressure gauge.
4) Increase engine speed steadily so that an oil pressure of 8 atm occurs.
5) Run engine up to power briefly and take-off. In this connection the temperatures of coolant and lubricant need not be kept under observation.

The time from firing to running) 1 1/2 mins at 0°C
engine up to power amounts) 2 1/2 mins at –10°C
approximately to) 3 1/2 mins at –20°C

With cold start operation avoid lengthy warming up of engine: thus start up only when take off follows immediately afterwards. If prolonged taxiing or idling cannot be avoided, see that the lubricant intake temperature does not exceed +40°C. (In summer +60°C).

Purpose of the Process

Owing to the viscosity of the cold lubricant the sliding parts of the engine tend to stick thus rendering the start difficult. In addition the lubricant pump can only suck in the viscous oil with difficulty and feed the various lubricating points of the engine.

The reduction of the viscosity hitherto effected by means of heat i.e. by warming up the lubricant or the engine can similarly be obtained by adding fuel to the lubricant without in any way affecting the lubricating quality of the latter. Since the fuel in the course of flight will evaporate from the lubricant with increasing heat of the engine, the lubricant will thus revert to its original state after a certain period of flying.

Thus the lubricant in the cold state (before starting up) as a result of the fuel addition and when warm (during flight) on account of the heat, flows freely and as before is effective as a lubricant.

Damage to the engine due to heat stresses as a result of taking engine up to power quickly does not occur.

J. Flight Readiness Report

First ground mechanic reports performing the following tests and inspections Nos. 1 – 31

Faults if any.

1) Locking devices of control surfaces free; all tarpaulins and covers removed.
2) Air pressure on three wheels 4.5 atm. Excess; no slipping of tyres (red marks)
3) All caps and covers on power plant and airframe secured.
4) Fuel and lubricant pipes tight.
5) Pressure oil pipes tight.
6) Coolant tank filled to the top.
7) 29.5 litres in lubricant tank.
8) Fuel tank filled to the brim.
9) Fuselage tail: Control surfaces, push rods, wire cables and roller chains undamaged.
10) Fuel hand pump functions.
11) Cabin firmly closed and bolted.
12) Pressure oil system filled (measuring (immersion) rod).
13) Oxygen breathing apparatus 150 atm. Excess.
14) Starting fuel in place.
15) Undercarriage switch in position “Rest”.
16) Flare pistol and ammunition installed.
17) Rudder, elevator and ailerons function smoothly and give proper deflections.

18) Throttle lever operates freely over whole range.

19) Fuel pressure pump functions properly.

20) Landing flaps ready for operation.

21) Both spoilers operate freely.

22) Generator running (starts functioning at approx. 1,900rpm)

23) Undercarriage brakes respond well.

24) When running engine up to power at full throttle:
Engine speed approx. 2,300rpm
Induction pressure approx. 1.35 atm. Abs.
Lubricant pressure 3.5 kg/cm2
Fuel pressure 1.3 –1.5 kg/cm2

26) Four engine attachments tight and secured.

27) All bearings for control surfaces are secured and locknuts tight.

28) Safety equipment in proper order.

29) Radio equipment inspected by radio mechanic.

30) Gun equipment tested by armourer.

31) Automatic system and hand adjustment of pitch variation of airscrew functions properly.

II. Flight Operation

A. Preparation for Take-Off

For starting up, warming up, running up engine, stopping and cold start preparation, see under "Getting ready for take-off".

1) Lock cabin securely (cabin handle in position 'closed'). See that the locking bolts snap in properly.

2) See that electric wire system is connected up.

3) See that main fuel cock is open.

4) When the air is extremely moist and the temperature below 0°C, switch on pilot tube heating through automatic switch on main switch panel. Indicator on instrument panel.

5) Switch on undercarriage position indicator; green lamps must light up.

6) Move all control surfaces for free action.

7) See whether automatic system for pitch variation of airscrew is switched on.

8) See whether spoilers are retracted (retract by pressing operating knob).

B. Taxying

1) Check whether temperature and pressure gauge register.
2) Landing flaps should be in zero position.
3) Open coolant and lubricant radiator louvers.
4) Tailplane to be in position 0°-1°
5) See that coolant temperature of 94°C is not exceeded.
6) For making sharp turns taxy straight for a short period at full throttle, then start turn with brake and throttle down.
7) When taxiing try out brakes. The brakes should respond well and should not lock wheels. Do not load the brakes too long on account of the heat generated.

C. Take-off

1) N.B. For cold start, observe "Cold Start Preparation" under Section 1. H.
2) Radiator and oil cooler flaps open.
3) Variable camber flaps in take-off position (20°). Adjustment wheel of flap on the left, indicator on left flap.
4) Tailplane in position 0° - 1°, adjustment wheel of flap on the left, mechanical indicator in front of the handwheel.
5) On take-off lubricant intake temperature of at least 30°C (press push button on instrument board).
6) The most favourable flight path speed for the climb is approx. 250km/h near the ground.
7) Retract undercarriage. Undercarriage switch to be in position "In". After retracting undercarriage the undercarriage switch springs back automatically into position "Rest". Should the switch not spring back, then pull at the switch handle.
8) Continuous indication of the position of undercarriage is obtained by means of the mechanical undercarriage indicator. Electrical indicating apparatus shows final position (red = completely retracted). Illuminated signs can be switched off; light up again automatically on lowering the variable camber flaps. Should the undercarriage not retract and lock completely, then put undercarriage switch again in position "in". If necessary hold undercarriage switch for a moment until electrical indicator lights up.
9) Put variable camber flaps into rearmost control position (external hand-wheel on left side wall of pilot's cabin). Indicator showing flap deflection on left flap. Trim (inner hand-wheel on left side wall). Both hand-wheels can be held simultaneously and operated.
10) Note: On failure of the fuel supply pump on the engine, it is not possible to obtain emergency fuel by operating the fuel hand-pump.

D. Flight

1) Flight limits.

Maximum permissible speed with variable camber flaps in operation and with flaps fully extended 250km/h.

Maximum permissible speed when operating undercarriage 350km/h.

Maximum permissible diving speed 750km/h.

Maximum permissible engine speed in the dive 3,000rpm (for 30 seconds only).

Best Indicated Speed Table for Climb

2)

Altitude m	0	1000	2000	3000	4000	5000	6000	7000
Speed km/h	250	243	236	229	222	215	208	200

3) Operational data

(a) Engine Speed and Induction Pressure (DB 601N)

Short period output	**Engine Speed rpm**	**Induction Pressure Atm. Abs.**
(5 minutes flight at full throttle)	2600	1.35
Boosted continuous rating (30 min)	2400	1.25
Normal Rating (cruising)	2300	1.15

Engine loading in patrol flight

Flying altitude	**Engine Speed rpm**	**Induction Pressure atm.abs.**	**Airscrew setting clock time**
1	1200	0.7	9.20
3	1200	0.7	8.30
5	1300	0.72	8.00
7	1500	0.60	8.45

For patrol flight switch off automatic blade setting of airscrew (tumbler switch on left fuselage wall) and set airscrew pitch by means of thumb switch.

Note: remember that the generator only starts functioning about 1,600rpm.

(b) Lubricant temperature

Lubricant intake temperature normal	60 – 70°C
Lubricant intake temperature maximum for short periods	80°C
Lubricant outlet temperature normal	90°C – 94°C
Lubricant outlet temperature maximum for short periods	105°C

(c) Lubricant pressure

Minimum 2.6 kg/cm2

In exceptional cases in the climb 2.4 kg/cm2

Maximum 5.0 kg/cm2

(d) Coolant temperature

Coolant intake temperature at ground level

Minimum 60°C

Maximum 75°C

Coolant outlet temperature (maximum temperature) for altitude.

Altitude	1-3	6	8	10
°C	95-90	85	80	75

Part III

Messerschmitt Me 109 G-2 Technical Description and Notes

Author's Note

The following reports are based on the Messerschmitt Me 109G-2, tropical version. The British captured this aircraft in December 1942 and RAF Middle East officers produced a description on which this is based.

Paul Sinnott, September 2000

Dimensions and Areas

Main Airframe Dimensions and Areas

Wing Span	32.6ft
Length	29.3ft
Undercarriage Track	6.5ft
Mean Wing Chord	5.7ft
Wing Area	172 sq ft
Total Flap Area	11.2 sq ft
Total Slot Area	6.7ft
Total Radiator Flap Area	8.6ft
Total Area of Ailerons	8.3 sq ft
Span of Tail Unit	9.8ft
Tailplane Area	15.7 sq ft
Elevator Area	9.25 sq ft
Fin Area	7.5 sq ft
Rudder Area	5.9 sq ft

Undercarriage

Undercarriage Leg	VDM 8-2787-05
Tyre Size	650 x 150 (Conti, Metzeler, Dunlop)
Tyre Pressure	64 lb/sq in

Pressure in oleo-leg (no load): 355 lb/sq in

Tail Wheel

Leg with helical spring and hydraulic damping

Tyre Size:	290 x 110
Tyre Pressure	64 lb/sq in

Propeller

Three-blade propeller, type VDM 9-12087 A.

Clockwise rotation as viewed from pilot's cockpit

Weight including blades, hub and v.p. mechanism: 375lb

Diameter	9.8ft

Engine

Daimler-Benz DB 605, B series. Inverted V 12-cylinder, liquid cooled.

Volumetric Compression Ratio:	7.3 – 7.5
Reduction Gear Ratios:	1 – 1.875

The maximum permissible engine speed in level flight is 2,800rpm and 3,000rpm during a dive.

Octane Fuel Rating used:	87

Engine Ratings

Rating	RPM	Boost		Altitude	Power	Fuel Consumption
		ata	lb/sq in	Ft	bhp	Gallon/hour
Take-Off	2,800	1.42	+5.5	SL	1,455	102.5
Climb (30 mins)	2,600	1.30	+3.8	SL	1,290	85.5
Max. Continuous	2,300	1.15	+1.7	SL	1,060	68.75
Max. Emergency	2,800	1.42	+5.5	18,700	1,335	93.75
Combat (30 min)	2,600	1.30	+3.8	19,000	1,235	81.5
Max. Continuous	2,300	1.15	+1.7	18,000	1,065	69.25
Economical	2,100	1.00	-0.5	18,700	880	54.25

Oil Cooler

Type – SKF/Behr FO 820

The cooling surface is 125 sq ft, capacity 0.9 gallon and the weight 60lb. The test pressure is given as 71 lb/sq in.

Coolant Radiators

Two light metal radiators, type SKF/Behr A1.F.750B.

The coolant is a mixture of water and glycol.

Cooling surface area:	125 sq ft
Capacity:	1.1 gal
Test Pressure:	28.4 lb/sq in

Header Tanks

Two light metal header tanks fitted, type NKF.

Capacity:	1.1 gal coolant, 1.1 gal air space
Total Weight:	7.3lb
Test Pressure:	17 lb/sq in

Oil Tanks

One light metal oil tanks, type NKF.

Oil Capacity:	8.1 gal plus additional air space of 1.3 gal
Weight:	12.75lb
Test Pressure:	5 lb/sq in

Fuel Tanks

One fuel tank manufacture by Karl Otto Raspe & Co.

Capacity:	88 gallons
Weight:	121lb
Test Pressure:	2.8 lb/sq in

Jettisonable Fuel Tank

One jettisonable fuel tank fitted, type Ju 87/NKF.

Capacity:	65 gallons
Weight:	55lb
Test Pressure:	2.8 sq in

Weights

Weights (lb)		**Fighter**		**Fighter-Bomber**	
	3 Guns no Drop Tank	2 additional wing guns	3 Guns and Drop Tank	With 1 x 250 kg bomb	With 5 x 50kg bombs
Empty	4,938	4,938	4,938	4,938	4,938
Flying	6,678	7,151	7,238	7,291	7,205
Maximum Flying Weight:		7,606			

Speeds

Engine Boost ata	Rating RPM	Altitude ft.	Max Speed mph
1.30	2,700	SL	305
1.30	2,700	23,500	375
1.30	2,700	30,000	364
1.42	2,800	SL	317
1.42	2,800	23,000	384
1.42	2,800	30,000	372

Climb

Engine Boost ata	Rating RPM	Altitude ft.	Rate of Climb ft/min
1.30	2,750	SL	3,350
1.30	2,750	21,500	3,400
1.30	2,750	30,000	1,400
1.42	2,800	SL	3,950
1.42	2,800	21,000	3,800

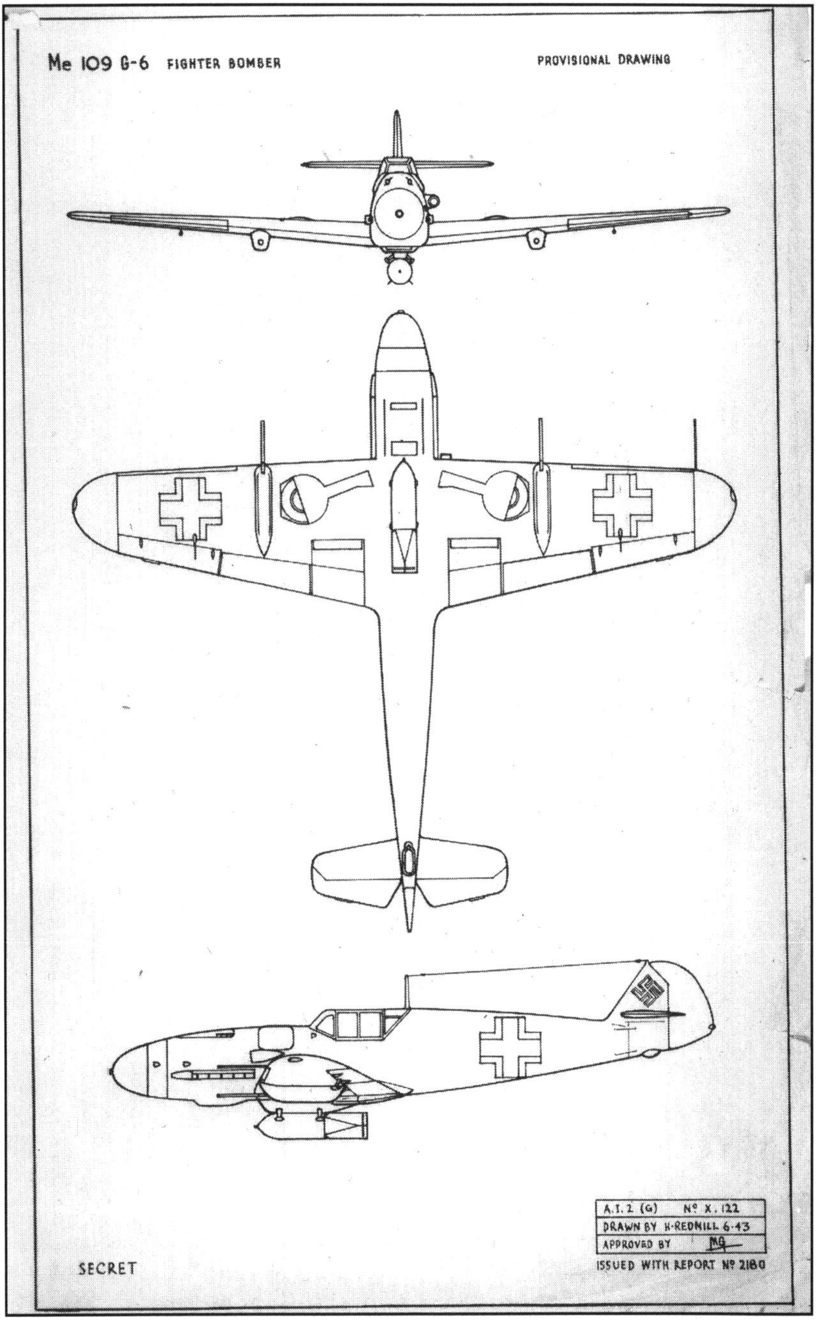
Me 109 G-6 FIGHTER BOMBER
PROVISIONAL DRAWING
A.I. 2 (G) Nº X. 122
DRAWN BY H-REDMILL 6-43
APPROVED BY
SECRET
ISSUED WITH REPORT Nº 2180

FIG. 3

. ME 109 COCKPIT